# Chromecast Device User Guide:

## Chromecast TV Device Setup and User Manual

By

Joseph Joyner

# Table of Contents

Chromecast Device User Guide: Chromecast TV Device Setup and User Manual

By Joseph Joyner

First Published, 2015

Printed in the United States of America

# Introduction

Since the internet speeds went above the 500kbps mark, the number of things people can do with their interconnected devices skyrocketed. Automated homes, connected cars, and internet enabled TVs. The latter is what threatens to revolutionize the entertainment world as we know it, do away with cable TV and abolish the need for expensive blue-ray players. Even though smart TVs are so lucrative, most people find them either too costly, or limiting. The alternative would be going for a TV with an HDMI port, a USB port and purchasing a dongle to bring in the power of internet TV.

Chromecast is one of the best and cheapest such tools in the market.

# Chapter 1. Chromecast Basics

It is barely two years old but it has managed to shake the industry in a way like no other. The staff at Google are apparently working their heads off to delivery us to the new world in terms of technology. Google Chromecast is a small Wi-Fi-enabled dongle device, which connects to a television through a HDMI port. It enables you to stream whatever it is you want to view on your TV screen from smartphone, laptop, tablet, or desktop. It was born in late July 2013. It is seen as a competitor to apple TV after the last task (Google TV) failed. It has an advantage over apple TV as it is comparatively cheap. It retails at $35 while Apple TV retails at $99. It also has a high audio and video quality.

This device resembles the USB thumb drive in every single way. It has a simple version of Google's chrome OS. Its memory is only 256k. It is not like it needs anymore than that amount of memory since it is not used for storing data. By installing it into your HDTV's HDMI port, you will be able to see the wonderful things that it can do. However, you will need a Wi-Fi network for the content to be streamed onto your TV screen.

The controlling tool in this case is your mobile device. You will be using the application son your phone to cast content on the Chromecast, which then gets the materials from the source in the internet and streams it on your TV. This enables you to continue using the phone for other things while streaming content on your TV. It also helps to save on the battery.

Chromecast can be used on a number of devices that include android enabled machines and chrome for Mac and Windows. The only exceptions are the blackberry and Windows phones which may just be for some time. To use your smartphone with the device, you must be in the same network to which the Chromecast is connected. In this way, you cast, or mirror your screen, videos or games to the TV.

By casting, it means that the entire screen of your device, or a specific app, will be forwarded to the big screen hosting the Chromecast dongle. This is especially fantastic since it cuts down on the number of cables trailing your room, from the devices to the TV. It also gives more flexibility meaning that you can always share whatever you are viewing with your friends at no extra effort.

The Google philosophy of being an open source entity has made it possible for many apps to work run on the device. They continue to increase with every passing day since as it is, the Chromecast is still growing. There are hardly any signs of slowing down.

Some of the most notable apps that run on the device include; YouTube, Hulu plus, Netflix, Pex media server Allcast, HBO GO, music apps, Google play, Vudu and crackle, and finally the chrome. With these apps, you can always bring your entertainment into life. You can either play games or stream HD video at all times. This takes you from the boundaries of cable TV that confined you to some of the subscribed channels and nothing more.

YouTube is the video app that houses any possible video dating way back and the present too. It is only natural that is is number one on our list. From music videos all the way to movies, you will be able to get it all. All the other apps are great in their own way. They have unique features which make them perfect for use with the Chromecast.

With the guest mode feature, family and friends can be able to cast onto your screen from their devices without having to connect to your Wi-Fi. This device has taken entertainment to another level. Your phone now has more better uses than calling, texting, receiving emails and checking out social networks.

# Chapter 2. Why You Need Chromecast

*It is affordable*

We are living in tough economical times. Anything that guarantees you saving a couple of dollars for exceptional services is welcomed by any means. It is just $35. It cannot possible get any better than that. It is better to make use of the offer as it still available. You never know, sometime along the line, the developers may just decide to raise the fee and sight a list of reasons you would not understand. Furthermore, the other viable option goes for $99. This would be that apple TV. At the moment, the Chromecast is your better option. Go for it.

*It hosts most of the apps that you would need*

At the beginning, Chromecast had only four compatible applications. These were YouTube, Netflix, Google play music, Google play movies and TV. With time, applications such as Hulu plus, Pandora radio, HBO GO joined in. This was a good move which later on lead to the Google team inviting app developers to a two day hackathon. The aim was to test drive the SDK's (software development kit) upcoming release.

Among those who attended there were 40 developers from 30 companies. It resulted in the production of 10 additional applications.

The SDK was later opened to all developers on February 3, 2014. What this means is that you should expect a lot more from chrome cast than what is available at the moment. It is still growing and being improved by the day. Getting your dongle today makes you part of the growth.

With an SDK in the market, the geeky ones could venture into the programmer in them to develop some clever app that will make their system more interesting.

*It is simple*

You only need the chrome cast (a small device measuring 2.83 inches) , your TV set, smartphone/pc, and Wi-Fi connection. That is all you need to get started on the journey to great entertainment. Installing will not take long. There is no remote required since your smartphone is your remote. It controls the volume and everything else. Provided you know your way around your phone, then you will not need any further training or understanding.

Its small size makes it possible for it to be carried around. You can therefore carry it to a friend's place to enjoy streaming movies, videos or music that you both love. The device also has low maintenance costs. After you have plugged it in, yours is to enjoy. Its power consumption is very low and does not use bandwidth when it is on standby.

*Offers high quality entertainment*

The dongle enables you to stream high quality HD pictures, videos and movies online. The internet is home to a lot of entertainment and Chromecast just gives you the ability to watch all of them on your television set. There are no regulations. Provided it is on the internet, you are able to view it on a large TV screen. If you happen to be one of those people who are tired and fed up with rules and regulations, it is time to move to a more liberal platform. You do not have to put up with boring television networks anymore.

# Chapter 3. Setting Up Chromecast

Once you have dished out the $35 and gotten your Chromecast dongle in exchange, setting up is as easy as it can get. Before you start the process of setting up the gadget, you need to ensure that you have the following gadgets: a display device that in this case would be a high definition television. It must have a HDMI input with access to a secure wireless network, and a USB port; connection to a Wi-Fi network; access to an available power network; and finally, a compatible iOS smartphone, tablet or Mac, Windows or Chromebook computer.

*Plugging it in*

The first thing to do when setting up the Chromecast dongle is to place it into the HDMI port on your HD TV. If you have another connector that you can use to extend the dongle it would be better. Doing so will increase the Wi-Fi range. Connect the TV to a power source. The dongle should also be connected to a power source too or if the TV set has a USB port; it can be connected to it.

*Setup*

The setup process is somehow similar for all the gadgets with only slight variations

*Install Google chrome browser*

This is the foundation of the whole process. The Google Chromecast will not work on any other browser. Your gadget has to have the browser before you can proceed.

*Download chrome cast utility*

Install the Chromecast application by going to chromecast.com then selecting setup. However, you will be required to specify the Chromecast you are downloading. This is because different Chromecast setups are meant for different machines.

*Launching*

After downloading, you will go ahead to launch the utility. It will start searching for the Chromecast device. Within a short time, it should be able to locate it whereby it will display the name of the device on the TV and the screen of your device. You must click "continue". A code will appear on the screen of your device and on the TV.

*Accept the code*

You should only click on "That's my code" if they match. At this moment, the internet connection of your device will be lost as the utility will have accessed your Wi-Fi connection. It will be configuring the Chromecast to connect directly to the Wi-Fi network.

*Confirm the wireless network*

With the ability to pull videos straight from the internet to your wireless network, the Chromecast will need to connect to your Wi-Fi. As a result you will have to specify the network that you will want the Chromecast to use. Enter the password and you will be given the option of changing the name of the Chromecast device. Once it connects, the screen of your TV will have heading that states "Ready to Cast". The same should appear on your phone together with a video that shows you how to use the Chromecast. By following the instructions you should be able to start enjoying favorite movies and videos immediately. It is as easy as operating your phone or pc.

## Chapter 4. Taking Your TV with You – Using Chromecast in Hotel, at Work

We have already seen that the Chromecast is a simple device that can be carried around easily. This makes your TV experience a portable adventure you can move around with anywhere you go. TV just became mobile. With the chrome cast, you only have to find a HD TV that has a HDMI port and internet connection and you are good to go. From hotel rooms to your work place, you are assured of the best entertainment that only you have control over.

For people who have busy schedules or are always travelling from one geographical region to the other, they won't have to worry about leaving their TV behind and missing out on their favorite programs. A stay in a hotel room is not going to be boring anymore. You will have your TV with you; all that would be missing is a screen. You connect it and you will have your home away from home with you.

Just like the invention of the plastic money, Chromecast has provided users with a very lucrative invention. Together with the cloud application, all that one has to do is store their favorite videos, music, and movies in the cloud and they can watch them from anywhere in the world. Streaming the videos from the cloud makes it safer and less hectic than having to download them and fill up your pc's hard drive.

By using Chromecast, you are not bound by geographical boundaries. You can easily stream channels from your home country when you are holed up in a far away land. Provided there is internet connectivity you can stay up to date with whatever is happening back home. You will hardly miss a thing. Even if you do, YouTube is filled with pre-recorded content that is usually uploaded occasionally.

The Google brand is available worldwide. It is loved by a good percentage of the world population and it blends well with most applications. This makes the Chromecast a better option compared to Apple TV. It is akin to having a universal credit card that is accepted in almost every country. Possessing one liberates you from the burden of carrying cash.

At the work place, you and your co-workers can catch up on the current affairs in the country by watching news. This is all thanks to the one gadget called

Chromecast. Going for a sleepover at a friend's place, your worry on the type of entertainment to engage in was answered when Chromecast came into the vicinity. This small gadget presents you with a whole world of entertainment for you to choose from. In case there are more than two guests, you will hardly miss something that entertains everybody. With Chromecast, you will be spoilt for choice. Your entertainment is right by your side everywhere you go.

Things in the entertainment realm shall not be the same again. It has all been laid out in your hands. You totally control your entertainment now. The phone is now the remote control.

# Chapter 5. How to Hold Teleconference with Chromecast

Teleconferencing is a very good way of communicating especially in the business world. Holding physical meetings sometimes is never possible when people are worlds apart. As a way of saving up on the air ticket if the two are to meet face to face. For many years, people have been able to use this method very effectively. With the entrance of the Chromecast into the vicinity, things became even better. The small pc screen can now be replaced by a large HD TV that can accommodate more than one person if the meeting is conducted by more than one person.

The good thing is that Chromecast is a Google product which runs well with the chrome browser which in turn supports the hangouts feature. This feature is already being used by many people. You can create a business hangout group for the company. This will be used purposefully for communicating between the workers and their bosses. It can serve as a good way of bridging the gap between the employees and their seniors especially if the two are located in different physical locations. Communicating orders across will be simple and cheap for the manager.

The company will need to invest in a large and state of the art television set. It has to be the best that the market has to offer if possible. The Chromecast is definitely a must have and a Wi-Fi connection. To avoid using the small camera on the pc or mobile phone, you can invest in a good webcam that will be placed strategically on top of the HD TV. With all that being ready, every time there is need for a meeting, you will only have to cast the video feed on the TV and proceed with the teleconference. Everybody will be visible and as a result communication will be clear.

This feature benefits the business in more ways than just cutting cost. The manager for instance is able to stay in touch with the happenings in his business without having to be physically present.

You should know that in every Chromecast in order to work properly you should have Wi-Fi with you, this is for doing some of your tweaks to the network and it is allowable as well. In some of the offices now, this is not considering as of their problem because many of the guests wants it and connects to their network.

# Chapter 6. 5 Things You Can Do with Chromecast at Work

1. Send any web page you can view in the Chrome browser to a TV. The potential here is huge. YouTube content, in it you can manage to work on them, and as well as all the stored data in your cloud– images spring to mind as likely candidates, as do your Google Drive documents. You could, for example, cast images of your latest products or pages of your website as you talk to potential clients.

2. Show your own video. Your company may well have video that it doesn't make publicly accessible through its website. That's not a problem; videos stored on your hard drive can be viewed in Chrome – if you press Ctrl+O, you can open video files in a Chrome tab, and then you can cast them. This simple way of casting videos means that you do not have to share essential clips before a presentation neither do you have to liaise with everyone before showing a funny clip. It makes life more fluid and easier to live or enjoy.

3. Broadcast a video conference. Another thing with chrome that is accessible is through the said Hangouts

– it has an in-app which you should be aware of is the Chromecast support like for the Skype. All developers allow in Google to tweak their products for Chromecast, and Skype is one of many ideal candidates. You can also cast your video conferencing to a TV – see the next tip...

4. Deliver traditional presentations from your laptop. Once you've set up the Chromecast extension on the Chrome browser you can mirror the display of your laptop. This means you can deliver presentations just by opening PowerPoint and running the presentation in it. This is a beta service for Chromecast and it is a bit hidden away. On your laptop click the Chromecast tab, you should see small arrow and then click to the right side of the screen of the word Beta. Should choose the "Cast entire screen (experimental)." It also has the support in audio.

5. Brainstorm. When you need to gather ideas and come up with strategies, do you still resort to a flip chart? For your table you should have the CastPad thats for the Chromecast if you are need of sending hand-drawn notes and as well as the images and put on to screen.

You can do an awful lot with just a Chromecast and the Google Cast extension for Chrome on your laptop. Things are ever changing so you should stay aware of for innovations that will let you make even better use of Chromecast in the office.

# Chapter 7. Additional Tips

*Watch whatever you want on big screen*

In this case we are talking about the restricted content that you will not find airing I your local TV channels. The world may be a global village but it is not too liberal. This is especially the case when you find that contents from one country are not allowed to be aired in others. A good example is the UK television show *The paradise* which US residents cannot be able to access. With the availability of Virtual Private Networks (VPN) and web proxies like media hint one is able to access the their favorite international TV shows. The web proxies and VPN enable one to get an Internet Protocol (IP) address in the country airing the content. Chromecast then makes it possible for you to view the program on your HD TV rather than your pc. The difference is quite big in terms of entertainment. You will also be able to view web content that can only be aired by use of specific applications like Hulu.

*Can be used in the place of the projector*

With the availability of the Chromecast, the projector would probably have no use at all. Compared to the two, the Chromecast has more advantages than the projector as it is simple, portable and can be put to more uses. For people who are engaged in public speaking, lecturers, or even students, there comes a time when you are required to make a presentation. The points have to be made visually for all to see. With the Chromecast you will only have to worry about a HD TV being in the room. You will be able to make your presentation easily and in a unique way.

*Play online games and watch own videos*

Playing online games just became more fun. With the ability of casting the game on a much bigger screen, you will be able to have more thrill playing the game than playing it on your pc. The images and characters are bigger which makes it more fun. Google has taken this factor into consideration and are now working on bringing games to Chromecast. A good example is the Tic Tac Toe which can be downloaded by the android and iOS gadgets. It is then cast on the TV and played.

The internet just made it possible to store videos and documents in the cloud. Accessing documents can be done through the phone or pc but for videos and music, you can easily play them on your TV by use of

Chromecast. This way, you will not have to worry about downloading the music, video and movie files before you can watch them.

*Video conferencing*

We have already looked at this one extensively above but it doubles up as a way through which you can have fun and be productive with Chromecast. By using the Chromecast and a web camera you can make it possible to hold a conference without having to squeeze to fit on a small pc screen.

*Share your tabs*

When running the Google Chrome browser, all you might need to reflect the things on your screen is a plugin that lets you forward or cast your tabs to the screen. With this power, it means you can use the screen to which the dongle is attached as a secondary screen without using a VGA or HDMI cable. For instance, you could choose to cast a single tab of your browser and easily share information with your collaborators. Alternatively, you could decide to work with casting the entire screen of your computer so that people can see what you are doing throughout.

Does this mean that you can replace your HDMI cables with a Chromecast? The answer lies in how you use your TV screen. If all you want is to forward video and are not concerned of any lags, then yes, you can. However, if you are into intense graphics oriented jobs, this would never be a good idea. The Chromecast will have a considerable amount of lag which can be affected further by the congestion of the Wi-Fi connection you choose to use in the process.

## Conclusion

Chromecast is an effective all rounded solution. If you want it for entertainment, it will keep you glued with a wide range of options. If you want to make your presentations at work from your tablet or smartphone, it brings your dream to reality. Teleconferencing on a bigger screen will be as simple as plugging the Chromecast into any TVs HDMI port. Combining this with the immense amount of support from developers gives you enough juice to keep your days rolling as you look forward to a more innovative tomorrow.

## Thank You Page

I want to personally thank you for reading my book. I hope you found information in this book useful and I would be very grateful if you could leave your honest review about this book.  I certainly want to thank you in advance for doing this.

If you have the time, you can check my other books too.

www.ingramcontent.com/pod-product-compliance
Lightning Source LLC
Chambersburg PA
CBHW071554080326
40690CB00056B/2033